VERTICAL

FARMING

How to combine business with environmental awareness

Gary Grending

INTRODUCTION

We all know that the Early Man was a cave-dwelling, hunter-gatherer. The hunter-gatherer lived a simple life. He thought that life meant his simple cave-home and hunting activities. Instead of worrying about the future, he focused gathering enough to feed him and his family for the day. However, as the human population grew, humans began looking for new ways to feed themselves.

The discovery of farming allowed humans to transition from a hunter-gatherer society into an agrarian society. Agriculture allowed humans to leave their caves and relocate to fertile regions.

The shift from meat to vegetables has kick-started an entire industry. The agriculture industry is considered the main livelihood for most of the people on this planet. Among farmers, soil-based farming is commonly conducted. This development, however, has changed our planet's map as well as the ecosystem. Regions where forests once stood, have been replaced with semi-functional areas. Agriculture is such a lucrative industry that some countries have entirely wiped out forests and replaced them with large plantations that deny any possibility of the re-existence of a natural ecosystem.

Large scale farming has helped tremendously in the population growth and development of nations. Throughout history experts have come up with new methods to enhance agricultural output. In fact, before the Industrial Revolution the Agricultural Revolution rocked Europe. The ideas that were brought forward at that time changed the profession completely.

The Industrial Revolution, however, has allowed scientists to come up with new methods that can increase Earth's fertility. These methods involve fertilizers; these are chemicals that enhance the soil fertility. Fertilizers have been quite effective. Previously, fertile land; like all living things, had a life. Today, however, the fertilizers have enabled the land to remain fertile for a longer time period.

The only drawback is the adverse effects these chemicals carry. Fertilizers have the ability to raise the soil's temperature, causing the soil to absorb more water than required. This is one of the main reasons behind global warming and decreasing water level.

Most experts believe that the population will exceed 8 billion people soon. This apparently means that in the coming years, there will be more mouths to feed.

We cannot grow more food If currently, global warming and mass deforestation are a major concern.

According to a basic calculation approximately 1 billion hectares are needed to feed such a large population. Although such a large arable land is not available. With the issue of global warming, we cannot cut down more trees, or inject the Earth with more fertilizers.

While farming has historically helped mankind grow and prosper, all our efforts have damaged the ecosystem. In order to farm new areas we have cut down trees that prevented soil erosion, irrigated land where there was a desert and carved farms in mountains.

This extensive interference by mankind has degraded the naturally fertile land and left it exposed for erosion.

Furthermore, the introduction of fertilizers and pesticides has polluted our water and damaged the freshwater aquatic life.

There is one possible solution though. The recently introduced method of building urban food production centers; or vertical farms, is a far more reliable method.

Vertical farms are indoor farms.

These can be built inside old warehouses, discarded shipping containers or inside abandoned mines. However, vertical farming is most commonly done in specially built skyscrapers.

Vertical farms can have a positive impact both; on the global population and the economy. Apparently vertical farms have the potential to generate enough food to feed one city.

The food produced on a vertical farm is free of fertilizers and does not carry any harmful agents that might cause consumers to fall sick.

While conventional farming is affected by many different factors (soil erosion, floods, torrential rains and urban migration) vertical farming is not. All of these factors have a direct impact on our food supply. At times, they even cause famines. Vertical farming, on the other hand, allows us to grow vegetables round-the-year, without any hassle. Apparently, vertical farms use food waste from restaurants and grey water as nutrients for the plants. Both these are available for the farms round-the-year, hence, no shortage of raw material. This also ensures that chemicals in the form of fertilizers are not used to enhance soil productivity.

VERTICAL FARMING

VERTICAL FARMING

Vertical farming can be defined as the execution of farming crops in vertically stacked layers.

Most of the vertical farming is done in old warehouses, skyscrapers or old shipping containers.

However, recently a number of entrepreneurs have started farming in specially built skyscrapers.

To understand vertical farming, one needs to expand the concept of greenhouse agriculture. Just like greenhouse agriculture, vertical farming too is conducted inside a building. However, where greenhouses limit themselves to producing garden plants and flowers, farmers can grow vegetables and medicinal crops in a vertical farm.

Vertical farming is conducted either in a hydroponic system or an aeroponic system. Hydroponics is the process of growing plants in water, while aeroponics requires plants to be sprayed with a nutrient solution. Both methods allow plants to grow without soil.

Citing growing demand for vertical farms, many manufacturers now produce devices that aid vertical farming.

LED lights are used by farms, along with natural light, to allow the plants to photosynthesis. Farms also use sprinklers to water the plants. These sprinklers aren't simply filled with water, they are instead filled with nutrient solutions.

However, the most important technology is the indoor environment controller. This sets the temperature, so that plants can grow healthy.

The Small island nation of Singapore, with a population of more than 5.5 million people, has a four story vertical farm that produces 1 ton of vegetables every day. Even though Singapore still relies heavily on imported food, that one vertical farm acts as a backup plan.

Because vertical farming technology helps farmers control each and every aspect of production, farms can grow food all year round. Precision farming helps farmers identify which plants need more light, water or humidity. By using the site specific crop management system they can reduce the work done by specifically providing care for the distressed plant, instead of the entire batch.

Another benefit is that vertical farms do not have any bugs. Hence, farmers won't have to spend on pesticides.

The capability to produce food indoors allows vertical farms to be located in or near the city. This can be used as an advantage by developed countries. Having an indoor farm near the city allows them to save costs incurred on transporting vegetables; from third world countries, stocking them in a warehouse and distributing within the cities.

Apparently, vertical farms allow metropolitan cities to grow all types of food except cattle, dairy, horses, and sheep; as these cannot be farmed at a vertical farm.

According to most agriculturists, around 300 square feet of indoor space is required to produce food for one person. Hence, a vertical farm would need as much land as one city block, covered with a building of as high as 30 stories, to produce food for 10,000 people. So, if conventional farms are replaced with 30-story high skyscrapers, we won't need to clear more land and build farms over it.

Many third world countries rely heavily on agriculture. These countries have replaced forests with farmlands.

The situation has become so intense that

large-scale deforestation has become a major concern in South America and Africa. In such a case, if vertical farming becomes the norm, it will prevent deforestation and the destruction of Earth's natural ecosystem.

One species that has been gravely affected by global warming, deforestation, building of dams and irrigation are the wild animals. Places like Africa, South America and Southeast Asia were previously, home to some exotic animals.

Today, unfortunately, several species are already extinct and several are near extinction. The only way natural habitat of wild animals can be preserved is by leaving the forests alone.

Vertical farms allow us to concentrate the farming activities near the cities. Hence, if the Government focus on establishing vertical farming enterprises the vast stretches of land will be left alone. And hopefully, without any human interference that land will return to its natural state.

CONTROLLED-ENVIRONMENTAL AGRICULTURE

The demand for food safety and sustainability has led agriculturists to come up with a method to grow food all year round.

Vertical farms use a technique called; "Controlled Environment Agriculture."

Controlled-environment agriculture is a scientific method commonly used in food production. The objective behind this is to provide ideal conditions for the crop to grow into a nutritious fruit or vegetable.

These ideal conditions can only be provided to the plant inside a confined location. This method uses technologies that can control the environment.

The CEA works well when the plant is grown in a soil-less system; mostly hydroponic, aeroponic, aquaculture and aquaponic systems.

Environment is one of the main factors that affects agriculture.

Farmers who conduct traditional farming are often worried that their crops might get affected by floods, or torrential rains.

The Controlled Environment Agriculture allows farmers to artificially set an environment that benefits all the plants in the building. This technology waters the plants through sprinklers that are placed over trays that carry the plants.

Unstable climate, lack of resources, pollution and soil erosion have damaged the agriculture industry and have affected the nutritional value of fruits and vegetables.

Controlled Environment Agriculture have the potential to secure and sustain food supply system. Allowing farmers to grow food without using chemicals to fertilize the plants. Additionally, the capability to produce food all year round has provided food companies to save on transport and logistics cost.

IMPLEMENTATION

Agriculturists have an option to choose either a fully environmentally controlled vertical farm system or an automated greenhouse with an artificially intelligent system.

The artificial greenhouse system can monitor the water supply, lighting, temperature, ventilation, carbon dioxide, and pests. It also has the ability to exclude contaminated nutrients from the hydroponics system, so that plants only receive fresh nutrient solutions.

The Controlled Environmental Agriculture can be adopted for any kind of crop. However, for such an expensive system to be installed, it is advisable to plant an economically viable crop. Apparently the system is very expensive. Since vertical farms depend on profits it is better if they grow cash crops. Cash crops; crops that have medicinal value, or edible algae can yield a good profit for the farm.

Some farms prefer growing salad vegetables; like lettuce, tomatoes, etc. These enterprises are known to produce thousands of tons of food in a year, which allows them to make a significant profit.

ECONOMICS

It is quite common to see dreamy young entrepreneurs come in this business and end up losing tons of money. Vertical farming is a technologically advanced business which carries immense potential. However, the equipment used is extremely expensive. Items like; LED lights, nutrient solutions, controlled heating systems are worth a lot. And it doesn't matter whether you implement the environmentally controlled closed-loop system, or the automated glasshouse system.

In order to make a profit, all the successful vertical farms are focused on growing salad vegetables or plants that carry medicinal value.

VERTICAL FARMING

HISTORY OF VERTICAL FARMING

It will be unfair to call vertical farming modern technology.

Foundations for vertical farming were laid long before the Industrial Revolution took place.

Apparently, the first recorded case of indoor farming dates back to Roman civilization.

The Roman emperor Tiberius was fond of the Armenian cucumbers. He would eat an Armenian cucumber every day. To fulfill the emperor's demand, royal gardeners were hired who used artificial methods to cultivate the cucumber daily.

According to records, many different nations have used vertical farms at different periods in time. The Koreans used a similar system to plant vegetables during the winter.

The Korean system involved the use of "ondol"; a Korean underfloor heating system to provide plants with artificially created warmth. This allowed the plants to flower and ripen in shorter time periods.

Similarly, Europeans have experimented with indoor farming for centuries. The Netherlands began using greenhouses in the 18th century. Its main aim was to grow enough fulfill the country's demand. Today, it has dozens of greenhouses that provide her with millions of fresh vegetables all year round.

Though so far, all of the attempts of indoor farming have been limited to greenhouses. It is true that some of the greenhouses are extremely large, and can produce several tons of food. However, none of these structures are multi-story buildings.

Greenhouses can only grow vegetables and flowers through artificial methods. However, indoor farming offers a wide range of food items.

Several architectural documents suggest that the idea of an indoor farming inside a vertical high rise was already there.

Prominent architects have worked on hydroponics, as early as the turn of the 20th century. In the early days, vertical farming was limited to greenhouses that produced flowering plants. Later, the greenhouse technology evolved into what is now known as vertical farming.

Sholto Douglas mentioned Armenian tower

hydroponics in his book; Hydroponics: The Bengal System. The book was written with data collected in the then East-Pakistan, present-day Bangladesh. Other books that mentions vertical farming is Ken Yeang's Bioclimatic Skyscraper and Pich-Aguilera's Garden Towers. Ken Yeang is best known for his ideas regarding mixed-use Bioclimatic Skyscraper. This skyscraper merges living units with food production.

One of the most well-known advocates of vertical farming in recent times was Dickson Despommier.

He was a professor of environmental health sciences and microbiology at Columbia University.

He re-introduced the concept of vertical farming in 1999. According to him, a 30 story farm on one city block was enough to provide food for 50,000 people. He said that if he grew hydroponic crops on the upper floor and fish and chicken on the lower floors, he would be able to provide a variety of options to consumers.

Although his ideas were challenged by environmentalists, he was able to popularize the concept. Many critics did have doubts regarding the high costs required to run the facility. However, Despommier's bold claims about producing tons of food for the entire city managed to get media attention. Several magazines; including the New York Magazine, covered the story. Several architects came up with ideas to build a skyscraper for the farm.

Since then the idea has gained widespread interest. Today several companies run vertical farms in Canada, Netherlands, Korea and the United States. The US Defense Advanced Research Projects Agency also has its own 18 floor tall farm, where it produces medicinal plants.

DIFFERENT TYPES OF VERTICAL FARMS

It is true that vertical farming comes in different shapes and sizes. Some farmers have used wall-mounted systems in their houses, while the more ambitious ones have turned large warehouses into a huge vegetable producing factory. However, there is one thing common in all of these farms i.e. all of these use one of the three soil-free systems; hydroponic, aeroponic and aquaponic.

HYDROPONICS

Hydroponics is a Latin word, which means "water working."

The hydroponics system can be classified as a subset of hydroculture. For those who don't know; hydroculture is a method used for growing plants without soil. This method uses water to supply plants with minerals.

When it comes to soil-based farming, the biological decomposition of soil breaks down the organic matter into basic nutrient salts that plants consume. The water present dissolves the salts, which are then consumed by the plant. In order for a plant to receive a diet everything, from water and soil to light and temperature, must have the right composition.

Vertical farming through hydroponics provides water that has become enriched with dissolved nutrients. Since the water in vertical farms is contained; unlike the water from conventional farming, it does not harm the environment in any way.

Vertical farms use a system that allows them to reuse water. This allows the farms; especially those in drought-hit areas, to be highly efficient in water usage.

HISTORY OF HYDROPONICS

Scientists, as early as the 17th century have been working on a method to grow plants without soil. According to those early scientific researchers, the growth of terrestrial plants in a soil-less environment, with the presence of nutrient solutions is called solution culture.

Today, solution culture is considered a form of hydroponics; without any inert medium. Notable scientist, William Frederick Gericke of the University of California at Berkeley was one of the early promoters of solution culture. In order to prove that the solution culture was good, he grew tomato vines at his residence. These vines were twenty-five feet high and had entirely grown in nutrient solutions. It was he who proposed the term hydroponics in 1937.

The earliest success of soil-less farming culture took place on Wake Island. Wake Island is a rocky atoll in the Pacific Ocean that is used by Pan American Airlines. Back in the 1930s, Pan America showed interest in growing vegetables there, for its passengers. It was apparently expensive to airlift vegetables from the mainland.

However, because there is no soil on Wake Island, hydroponics was the only option available.

Recently, the National Aeronautical and Space Administration (NASA) has conducted considerable research on one of its projects; Controlled Ecological Life Support System (CELSS). Hydroponics culture in a way copies the Martian environment when it uses the LED lighting to supply heat to the plants.

A researcher at the Kennedy Space Center's

Space Life Science Lab believes that hydroponics can provide astronauts a regenerative life support system.

NASA has been researching a bioregenerative life support system for quite some time. It believes that with this system, NASA can send astronauts to Mars for much longer periods, without worrying about food supply.

Today, Canada is heavily invested in hydroponics farming.

Currently, it has allowed many entrepreneurs to turn hundreds of acres of land into large-scale hydroponic greenhouses.

These farms are currently growing commonly used vegetables; tomatoes, peppers, lettuces, and cucumbers. With more technological advancements, this industry can grow to $700 million by 2023.

SUBSTRATES

A hydroponic system might be soil-free but still needs a medium(substrate) to anchor the plant roots. A substrate can be sand, granular or pebbles. All of these mediums are porous and allow air and water to reach the roots. apparently , roots must have adequate access to air and water, as they need to breathe as well. Additionally, a porous medium allows nutrients to be delivered directly to the plant's roots. This helps, since plants in a vertical farm are stacked in shelves, they don't have enough spaces to let their roots spread out. The space saved can be used to plant more plants.

- **SANDY SUBSTRATE**

A sandy substrate has particles with a diameter of 0.6 mm to 2 mm. Coarse sands; 2mm diameter, absorb a lot of water and are not a good choice for hydroponic farming. On the other hand, the small-sized particles can clog the pump of the hydroponic system. However, the finer sand particles are cheap and readily available. Additionally, when wet, fine sand can provide the plant roots an anchor.

Sand also contains some absorbable nutrients. These nutrients in sand culture differ significantly in accordance with the substrate's color and origin.

Usually sand has large quantities of shell fragments which result in high calcium content.

Mostly black sand has a strong magnetite content that originates from volcanic rock.

According to agriculturists, black sand has qualities that make it fertile.

Yellow and orange sand colors have high iron content as indicated by their color.

White cells are known to have high silica content that can help plants build healthy cell walls.

- **GRAVEL**

Granular particles have diameters between 2 to 4 mm. They contain either plant mulch or gravel. Usually, hydroponic farmers use stone gravel in hydroponic systems as they form substantial, non-biodegradable anchors for plant roots. There is a variety of stone gravel to choose from, however, unlike sand, they hardly have any plant nutrients.

- **CREEK ROCK**

Creek rocks consist of round stones that are naturally cut very fine, giving them a shine. The natural smooth shape of these stones allows air to pass through and reach the plant roots. Crushed rock, on the other hand, has sharper edges and can easily interlock. This ability allows them to retain more water within the hydroponic tank.

- **STONE-BASED SUBSTRATE**

Stone-based substrates have the advantage of being reusable. Apparently, because when boiled, they are less messy than the sand substrate. If the desired plant is not expected to reach a considerable height than you can use plant mulch. Mulches are known to retain a large amount of water, but they also breathe well. Hence, plant mulch will allow air to reach the plant.

However, the stone-based substrate is highly

degradable and can clog pumps.

Simultaneously, the growth of mold and algae pose a high risk when mulches are used. The only advantage they have is that they do not have to be reused. Hence, once used, stone-based substrates can be replaced with fresh material. If you use the hydroponic system during the offseason then you might find this convenient, though.

- **PEBBLED SUBSTRATE**

Pebbled substrate range between 4mm and 64mm. Stone pebbles have the same qualities that creek rock has. They are smooth, shiny and the way they interlock allows water and air to pass through. However, only stones with pockmarked or matte surfaces are porous and allow perfect aeration. The stones that have shiny surfaces provide minimum water retention.

Make sure not to sterilize the pebbled substrate as they can explode when heated.

- **EXPANDED CLAY AGGREGATE**

Expanded clay aggregate is basically baked clay pellets.

These are a good substrate as they allow hydroponic systems to control the water solution. Additionally, the clay pellets do not contain any nutrient value of their own. They are pH-neutral and inert. Clay is shaped into round pellets and fired in a rotary kiln at 1200 °C, resulting in the clay to expand and become porous. The pellets are lightweight and do not compact, however, their shape might vary, depending on the manufacturing process.

Most manufacturers argue that clay pellets are ecologically sustainable and can be reused after it is sterilized. They believe that by washing the pellets in a solution of white vinegar, hydrogen peroxide and chlorine bleach the pellets can easily be cleaned of waste.

On the other hand, some manufacturers think that since the plant roots have broken in the clay pellets, it is best not to reuse them.

- **GROWSTONES**

Growstones is a substrate made from glass waste. It is said to have more retention space for air and water than peat and perlite. This substrate can contain more water than half-cooked rice hulls. Calcium carbonate has a presence with a volume of 0.5 to 5%. In other words, for every 5 kg bag of Growstones, there are 258 grams of calcium carbonate. While the rest is soda-lime glass.

- **COIR PEAT**

Coco peat or coir peat, is basically what is left after the fibers are removed from an outer shell of a coconut. Coir is a naturally grown flowering medium. Coconut coir has Trichoderma fungi, that secures plant roots and triggers root growth. It is quite hard to over-water the coir peat because of its excellent air-water ratio.

Apparently, this is a hydroponic friendly substrate. Coirs also have high cation exchange. This means that it can cache unused minerals and release them whenever required by the plant.

- **RICE HUSKS**

Half cooked rice husks have little use. However, they can be used by hydroponic farms once they start decaying. They retain less water, though this has never impacted plant growth.

- **ROCK WOOL**

Made from molten rock, rock wool is the most common substrate used in hydroponic farms. Apparently it is its chemical composition which allows it to be protected from common microbiological degradation.

It is an inert substrate that favors both; recirculating and run-to-waste systems.

Rock wool has many advantages, as well as disadvantages.

One disadvantage can be the probable skin irritancy when it touched with bare hands.

Another disadvantage includes the high pH value of mineral wool which makes it quite unsuitable for plant growth.

However, some conditioning can allow it to

be used by the farms.

A good advantage is proven effectiveness as a common hydroponic substrate.

Rock wool can be manufactured in a way to hold a large amount of air and water. This helps nutrient uptake and root growth in hydroponics.

- **POLYSTYRENE PACKING PEANUTS**

Polystyrene packing peanuts are cheap, easily available and have remarkable drainage. Though, they might be too light for various uses. They are often used within a closed tube system. However, users need to vary that only non-biodegradable polystyrene peanuts should be used, as biodegradable polystyrene peanuts decompose while usage and end up becoming a sludge.

Additionally, if a plant absorbs the styrene and passes it on to its consumers than there might be a health risk.

NUTRIENT SOLUTIONS

- **INORGANIC HYDROPONIC SOLUTIONS**

The idea behind nutrient solutions is to remove soil. This concept is especially beneficial for those people who live in places where fertile soil is not as commonly available.

Nutrient solutions are a mixture that contains just as many nutrients as the soil has. However, there are many ways that a nutrient solution can differ from soil chemistry.

For instance, in contrast to soil, nutrient solutions don't have cation exchange capacity, present in organic matter. This means that nutrient concentrations and pH value can change much more quickly than in soil-based setups.

If the pH value is not adequate for the hydroponic setup or the water is contaminated, iron can precipitate. Resulting in plants not receiving any iron at all. To avoid this, pH needs to be adjusted routinely or chelating agents have to be used for the plant to receive iron.

Routine adjustments should not be considered as a disadvantage of hydroponic farming.

Apparently, in accordance with Liebig's law, routine adjustments are a part of conventional farming as well. For farmer's ease, nutrient solutions that have an acceptable amount of concentration. Some of these solutions can be used for more than one plant type. Most often nutrient solutions have concentrations ranging from 1000 to 2500 ppm. Concentrations that are below these levels might lead to nutrient deficiencies, while those with higher levels can result in nutrient toxicity.

- **ORGANIC HYDROPONICS**

In conventional hydroponics, organic fertilizers are a good replacement for inorganic compounds.

Though organic compounds come with a baggage full of difficult challenges. Organic compounds have extremely variable quality, owing to the diet the animals have consumed. This not only varies the quality of the product but can also contain diseases, which will harm the end-user.

Additionally, organic fertilizers most often clog the system, resulting in extra, unnecessary work for the farm. And if the material clogged degrades, the hydroponic farm will end up stinking for hours, or possibly days.

VERTICAL FARMING

AQUAPONICS

An aquaponic system is built on the hydroponic system, as it combines plant farms and fish farms together in one ecosystem.

Fish are raised in vertical farms and their waste acts as a form of nutrient for the plants. Plants reciprocate by acting as a natural recycler when they purify the wastewater.

To understand this better, the aquaponic system allows water to be fed to the hydroponic system allowing the Nitrifying bacteria to break down the by-products into nitrates. The purified water is then recirculated back into aquaculture.

HISTORY

Most people have learned about aquaponics recently, however, you might be surprised to know that aquaponics is a centuries old practice.

In ancient times, the Aztecs used to use this technique to cultivate their farm islands. They used to call these farm islands, 'Chinampas.'

Chinampas required plants to be grown on either stationary or movable islands. These islands were constructed in lake shallows.

Waste material dredged from the chinampas canals and nearby settlements were used for irrigation purposes.

Southeast Asia and China are known for large-scale rice farming. Since rice paddies have to be submerged in water, the farmers would raise fish along with rice. This practice is the closest to the modern-day aquaponics system.

Recently China has introduced a floating aquaponics system on fish ponds. This allows the Chinese farmers to grow rice, canna lily, and wheat on an area exceeding 2.5 acres.

The New Alchemy Institute deserves credit for modern aquaponics. Inspired by the work done at the New Alchemy, other institutes have also conducted research on this topic.

Most notable among these is the research conducted by Dr. James Rakocy of the University of the Virgin Islands. Dr. Rakocy developed the application of a large scale, deep water hydroponic grow beds.

Studies of the system have shown that the system can run efficiently at low pH levels, which is favorable for plants.

COMPONENTS OF AN AQUAPONICS SYSTEM

The aquaponics system is made up of two main parts. The first part is aquaculture, which raises aquatic animals. The second part is the hydroponics part, which deals with plants grown in a soil-less environment. Aquatic animal waste collects in the water tank and acts as a nutrient solution for the plants. This effluent rich water might be toxic for the fish, but it is necessary for the plants.

On the front, the aquaponics system consists of these two parts. However, in reality, it has several components working together at the same time. Some of these components are;

- **Rearing Tank:** This is the component where fish are raised.
- **Setting Basin**: Uneaten fish food and disconnected biofilms and any other particles that are not needed, collect here.
- **Biofilter**: This is the area where the nitrification bacteria grows. It later converts ammonia into nitrates, which can be used by the plants.
- **Hydroponics Subsystem**: Plants are grown in this portion when they absorb nutrients from the water.

- **Sump**: This is the lowest point of the system. The water flows to this point and then it is pumped back into the rearing tanks.

In order to reduce costs and improve the functioning of the system, the units are combined into one big unit. This prevents the water from flowing directly from one part to the other end.

LIVE COMPONENTS OF THE AQUAPONICS SYSTEM

The aquaponic system does not only consist of machines, but it also has live components. The following are the three components which make up the system;

- **PLANTS**

Most plants can be grown in an aquaponic system, although plant growth depends on the aquatic animals.

The aquatic animals are the primary reason behind the concentration of nutrients and their availability for the plants. Plants such as lettuce, Chinese cabbage, basil, spinach, herbs, chives, and watercress can easily adapt to an aquaponic system, as they require a medium amount of nutrients. While plants like tomatoes, peppers, and cucumbers require a higher amount of nutrients and hence, they will fare well in a system that has a high stocking volume of fish. Following are some more plants that can be grown;

- **SEAWEED**

Edible Algae

Edible seaweed is a type of algae that is commonly used in food. It contains a high amount of fiber and protein. Edible seaweeds are believed to be one of the many groups of multicellular algae.

- **Carrageenan**

The Carrageenan comes from a group of linear sulfated polysaccharides that are taken from red edible seaweeds. There is a lot of demand for the Carrageenan, especially for their gelling, stabilizing and thickening properties. Mostly they are used to bind the meat and dairy products.

- **FISH**

Freshwater fish have the ability to tolerate crowded spaces, hence they are ideal for aquaponics. However, other practices involve saltwater aquaponics, coldwater aquaponics, and warm water aquaponics. It totally depends on the type of plants you want to grow and the resources you have available.

Most aquaponic systems use tilapia, eel-tailed catfish, silver perch, barramundi, and

Murray cord. For places where maintaining water, the temperature is a problem, catfish and bluegill are the best options.

- **BACTERIA**

One of the most important parts of an aquaponic system is the aerobic conversion the ammonia gas into nitrates, also known as Nitrification.

This process is important for the system as it decreases the toxicity of the water and allows the plants to remove the nitrate particles. Ammonia must be removed immediately as a high concentration of ammonia causes damage to tissues and might even kill the fish.

ECONOMICS OF AQUAPONICS

Aquaponics is still grown in smaller farms, as most large vertical farms don't invest in this system. They feel that fast-growing vegetables are more profitable than fish. However, if cheaper models of aquaponics are introduced there is a high chance that it might be adopted by more farms.

Aquaponics is a viable option for small farms because it offers farmers two sources to earn profit from. Additionally, consumers who prefer to eat healthy food should know that both of these products are organic and free of pesticides.

AEROPONICS

The aeroponics system involves growing farms in a mist or air environment, without using any substrate. The aeroponics system is quite different from both of the previous two systems. Unlike hydroponics or aquaponics, where a medium is required to hold the plant down and water is used to supply nutrients, Aeroponics does not require a medium. Because of its sensitivity, aeroponics is usually combined with hydroponics and acts as a backup system.

HISTORY

Scientists have been researching Aeroponics since the beginning of the 20th century. This concept, however, was first introduced by V. M. Artsikhovski, back in 1911.

In his paper, he talked at length about a soil-less agricultural system that can prosper simply by spraying the exposed root system with a nutrient solution. He claimed that such a system will be quite suitable for cultivation.

It was in 1957, when F. W, Went named the process as 'aeroponics'.

Since that time many researchers have tried and succeeded in growing a number of plants in an aeroponic system. Citrus plants, avocado roots, apple trees, coffee plants, and tomatoes have all been grown by various researchers. Today, aeroponics is practiced all over the globe, and several companies that manufacture aeroponics equipment.

- **GENESIS MACHINE**

In 1983 a company, known as GTi launched aeroponic equipment. The machine they launched became known as Genesis Machine; inspired by Star Trek II: Wrath of Khan. This machine consisted of an open-loop water-powered system that was controlled by a microchip. This machine was connected to an electrical socket and a water faucet. It produced a hydro-atomized, high-pressure nutrient spray within the aeroponic chamber, allowing the machine to water the plant roots.

- **AEROPONIC CLONING**

Had it not been for aeroponics, plant cloning would have never taken off. Hence, it won't be wrong if I claim that aeroponics is responsible for the modern changes in plant cloning. Apparently, a fully automated aeroponics farm allows the farmers to use *'cutting,'* for the plant cloning. This method has helped the widespread cultivation of cacti and hardwood, which were previously not easy to clone. In other more conventional agricultural cultures, plant cloning takes time and a lot of effort. However, thanks to automation, aeroponics culture allows farmers to grow hundreds of thousands of plants with minimal effort and resources applied.

- **GENESIS GROWING SYSTEM**

GTi, the manufacturer of aeroponics systems, was quite active in the 80s. However, it didn't stop with one technology. In 1985 GTi launched another aeroponics hardware. This new piece of equipment was a closed-loop system which can recycled effluents for reuse. This new system was called Genesis Growing System. With this technology, the company sealed its fate as a leader in the aeroponics system.

- **SPACE AEROPONICS**

People have always wondered what astronauts eat when they are in space. Well back in the 60s, when the first space shuttle went in space astronauts used to eat canned food. However, the initial space flights and astronaut's overall tenure was shorter than it is today. Today, astronauts spend nearly a year in a single flight, at the International Space Station.

In order for NASA to supply its astronauts with a food supply that will last throughout the year, it has to come up with new ideas. And what better idea than aeroponically growing food in space.

Since the first space flights, astronauts have been transporting plant seeds with them to research the effect of the space environment on plants. So far seeds of wheat, maize, pea, and spring onions have been carried into space. Later NASA and other research organizations performed several experiments on various plants. These experiments were conducted on Earth, in a controlled, low-gravity environment.

These experiments were conducted on different plant species like; pine, mung bean, cress, oat, and lettuce. The plant reactions included decreased seedling and root and shoot growth in low-gravity. Additionally, some plants; including peas, showed a high level of potassium and phosphorus when grown in space. At the same time peas also showed a decrease in zinc, iron, magnesium, manganese, and calcium.

In order to increase the fertility of aeroponic plants, NASA introduced a biocontrol liquid called 'Organic Disease Control'. This liquid controls pathogens in a closed-loop aeroponics system, without any pesticides. The initial success of Organic Disease Control (ODC) has lead NASA to use it in more experiments, on a variety of plants. The advanced experiments conducted throughout the late 90s resulted in ODC be improved. It is now the go-to substance for pesticide-free aeroponic, as well as organic farming.

METHODOLOGY

The basic idea behind the aeroponic system is to suspend plants in a semi-closed or closed environment and supply it with nutrients by spraying the plant roots and lower stem with a spray. Usually, the labor and expenses are reduced by squeezing the closed-cell foam around the lower stem and placed into an opening.

Theoretically, the aeroponic system should have a controlled environment, free from pests and diseases. However, because the plants are exposed, not inside a tank with a medium holding them, they are prone to get affected by pests and diseases. While the plant's canopy and leaves extend above, the plant support structure separates the roots.

ADVANTAGES OF AEROPONIC SYSTEM

- **MORE EXPOSURE TO AIR**

The idea behind an aeroponic system is that plants require access to air for better growth. However, there is a strong possibility that aeroponic plants can be affected by diseases. Farmers should take extra care not to allow devices, materials or people with diseases to enter the area.

- **IMPORTANCE OF OXYGEN FOR PLANT ROOTS**

Oxygen is vital for plant roots. Apparently, oxygen helps plants grow far better. Hence, the idea behind aeroponics is to feed plants with as much oxygen as possible. Most farmers believe that the increased aeration in the aeroponics system helps stop pathogen formation. They claim that clean air performs as a purifier and helps prevent plants from getting infected by diseases. Simultaneously, researchers have found that plant roots have grown more and better in an aeroponics system.

- **CO2 IN AIR BOOSTS GROWTH**

Aeroponic plants have access to a lot more CO2 than those in other systems. CO2 concentrations range from 450 ppm to 780 ppm, depending on how far above the farm is from sea level. Usually, lower elevations have a high concentration of CO2.

Studies prove that CO2 is good for plants.

A high concentration of carbon dioxide impact plants in different ways. First, they improve crop yield when they raise the rate of photosynthesis, which prompts growth. Secondly, it decreases the amount of water, plants lose because of transpiration.

Plants lose water through the small pores on their leaves. These pores; called Stomata, collect CO_2 molecules for photosynthesis while releasing water vapors. However, when carbon dioxide concentration increases, the pore opening becomes limited; they don't open wide enough. This results in a decrease in plant transpiration and more water conservation.

- **ASEPTIC CULTIVATION**

The main reason why aeroponics or any other form of soil-less farming was introduced, is the fact that soil can get infected by chemicals, used by factories, which in turn infects plants, with diseases. The Aeroponic system allows farmers to keep plants at a distance from other plants, and remove diseased plants immediately. This measure prevents aeroponic plants from being infected and helps them grow at a higher density than those grown through conventional methods.

- **RESEARCH VEHICLE**

Aeroponics has become an important field for researchers. Many researchers conducting research in botany, with emphasis on root morphology, find it easier to study roots in an aeroponic system.

The ability to control the amount of water plant roots require has allowed scientists to discover farming methods that require far less water than usual. This new discovery can allow the scientific community to address food shortages in drought-hit areas. So far scientists have found out that deserts and other unfertile regions can benefit from aeroponics. These regions can use the aeroponic system to produce their own crops, instead of importing from the third world countries.

TYPES OF AEROPONICS FARMING

- **LOW-PRESSURE COMPONENTS**

There are low-pressure farms with aeroponic systems. The plants at these systems are either hung above a pool of nutrient solution or inside a tube which connects the plants with the pool. Low-pressure pumps supply the nutrient solution to the plants through jets. The water in low-pressure components has to be adequately sprayed on all sections of the root system.

Contrarily, a dry section might prevent the required nutrients from reaching all parts of the plant. Hence, the farmers should make sure that the water jets are strong enough to reach every portion of the root.

- **HIGH-PRESSURE EQUIPMENT**

Aeroponics system requires high-pressure

equipment to supply water to all parts of the roots system. Additionally, a high-pressure system also functions as an air and water purifier. This equipment may sound expensive, however, if you compare costs with the output you receive you'll find the prices comparatively reasonable.

OTHER FORMS OF VERTICAL FARMS

- **SKYSCRAPER VERTICAL FARMS**

For mass scale vertical farming some ambitious people have experimented with skyscrapers.

This idea was first introduced by Ken Yeang. He believes that vertical farms should not be sealed. Instead, they should be open-air structures, with a controlled climate. Hence, he designed a mixed-use skyscraper.

Yeang's version of a vertical farm is better suited for personal or small-scale use, rather than a large-scale production which feeds an entire town.

However, ecologist Dickson Despommier proposed that skyscrapers can help the countryside to revert to its natural system. He believes that mankind's goal of large-scale agriculture has damaged the ecological system through deforestation.

Our efforts have resulted in global warming, natural disasters, large-scale desertification, diseases, and drought. Hence, by building skyscraper vertical farms we will be forcing the common population to leave the countryside alone, and allow the Earth to heal.

- **VERTICAL FARMS IN SHIPPING CONTAINERS**

When it comes to vertical farming many people use their imagination. For common people, a vertical farm can be a wall in their home.

For small-scale businesspersons, it can be a shed.

For a millionaire, it can be a skyscraper, and for NASA a vertical farm can be built at the International Space Station; in the space.

If you use your imaginations, you can build a vertical farm anywhere. You might need to invest your hard-earned money. But how much depends on the size of your market.

Some companies have found that building vertical farms in used shipping containers has two benefits,

a) You will be recycling the container,

b) your vertical farm will be cheaper than your competitors.

Containers provide standardized environmental chambers which are ideal for growing plants.

• DEEP FARMING

Some ambitious farmers have gone so far as to build vertical farms inside abandoned mines.

Although dangerous, deep farming allows plants to take advantage of underground temperatures. Besides, the farmers don't have to spend money on building premises, because they already have an abandoned mine, that can be used.

HOW VERTICAL FARMING WORKS

Vertical farming is a technically advanced form of agriculture.

There are far more technological aspects to a vertical farm than there are for a conventional farm. In the previous chapter, we discussed the various methods used in vertical farming, in this chapter we will highlight the equipment used to build a vertical farm.

Unlike conventional farming, a vertical farm has a very different setup. It comprises of the following four essential elements, without which it would be impossible to conduct it.

- **PHYSICAL LAYOUT**

The main objective of setting a vertical farm is to produce more food. Hence, in order to achieve this objective farmers have to plant crops in stacked layers, in a building. In a vertical farm, your output is measured in terms of output efficiency per square meter. Hence, in order for you to achieve this, you will have to build or buy a very tall structure that can hold plant trays. The higher the structure, the more food you grow per square meter.

- **LIGHTING**

Since the plants are inside the vertical farm, and not out in the open, the farm have to provide adequate lighting to the plants. This light is used by the plants for photosynthesis. At most vertical farms a combination of artificial and natural light is applied so that plants can attain maximum energy. Specially designed technology, such as the rotating beds are used to increase efficiency.

- **MEDIUM**

Plants need a medium, which will support them, and act as a transporter of nutrients. In conventional farming, soil acts as a medium. Soil allows support for the plants. In places where there are many plants, plants use soil to transfer nutrients from one to another. In return plants spread their roots through the soil and hold on to it, preventing soil erosion. Vertical farming, however, has three different versions and each version has its own idea of a medium.

❖ HYDROPONICS:

In hydroponics, plants are grown in water. This water contains nutrient solutions, through which nutrients are transferred directly into the plant. Experts of hydroponics claim that they do not need to change the water every now and then. Instead used water is recirculated.

❖ AQUAPONICS:

Aquaponics is a bit more advanced version of hydroponics. It is a closed-loop system that allows farmers to grow both; plants and fish.

Farmers who live in areas where fish consumption is high have managed to benefit from aquaponics culture. In aquaponics, the fish supply the plants with nutrients and bacteria. In return, the plants purify the water. Hence, the medium for aquaponics is the fish-filled water.

❖ AEROPONICS:

Aeroponics farming is totally different from both the previous methods. Aeroponics allows plants to grow in mid-air, without the need of any medium. The plants receive nutrients when the farmers spray the nutrient solution on their roots. The system is more cost-effective than the previous ones, as it does not require any medium to take care of. It also allows scalability, as plants do not have to be unearthed to have changes made to their roots.

• SUSTAINABILITY

Vertical farming has a lot of potential for technological advancement.

The interesting thing is that many companies have already built advanced equipment which allows sustainability in vertical farming. Some of this equipment has allowed farm owners to save energy, water, and money.

All three of these things are used abundantly in conventional farming.

- **WATER**

Plants require water, whether they are grown on a conventional farm or at a vertical farm. However, vertical farming is conducted in ways that conserve water.

The required nutrients are mixed in water and supplied to plants. The used water, instead of being dumped, is re-used.

Hence, vertical farming is a blessing in disguise for water-starved countries.

PROS AND CONS OF VERTICAL FARMING

PROS

Even though vertical farming is not a new concept, it has still not taken over the agricultural industry.

Apparently, most countries still don't consider it as a possible solution to food shortage.

The few countries that have taken vertical farming seriously are already producing hundreds of thousands of vegetables.

Vertical farming is profitable for only those who scale up production.

Food production is worth billions of dollars. And besides the monetary value of food, it is vital for mankind's survival. Hence, companies involved in commercial vertical farming are the only ones who will benefit from this venture. The following points explain what advantages vertical farming provides.

- **DOES NOT REQUIRE A FARMLAND**

Currently, the global population is increasing so rapidly that in a few years we may face a housing shortage.

Imagine, in such a situation we would have to cut down forests, possibly farms too, and expand cities.

Already some third world countries are going through this phase. To prepare for such a situation, it is better if we build vertical farms. Especially because vertical farms are usually built in urban areas. Vertical farms have stacks of plants shelved, one over the other which allows them to cover less space.

- **VERTICAL FARMING BOOSTS CROP PRODUCTION**

Contrary to conventional farming methods, vertical farming does not depend on soil fertility or the weather. Vertical farmers have the technological ability to produce crops round-the-year. Vertical farms allow farm owners to grow the same amount of crops in which a traditional farmer would grow on a large farming estate, by stacking plant trays one over the other.

Apparently, a multi-storeyed skyscraper has the ability to replace traditional farms.

Year-long farming increases crop productivity and allows cities to import fewer crops from outside. Additionally, since vertical farming is done in cities, therefore costs of transportation, warehousing, electricity (for the refrigeration) and pesticides (to avoid infestation) can be avoided. With reduced costs and increased productivity, vertical farming is more beneficial for modern cities, than importing food from across the world.

- **LESS DEPENDENT ON WEATHER**

Conventional farming is heavily dependent on the weather. Weather is the one thing that we humans have still not been able to control. Hence, countries with extreme weathers; like Canada, Saudi Arabia, etc. usually have to rely on imported food. Even agricultural countries are in constant fear of their crops being affected by a possible change in weather.

Plants in a vertical farm grow inside a building. The only environmental change these farms are affected by earthquakes and tornadoes. The environment inside the farm is set up by the vertical farm management themselves. Hence, there is no chance of rainfall, snowstorm or a wildfire wiping out this year's harvest inside a vertical farm.

- **LAND CONSERVATION**

At the core of vertical farming is the concept of conservation.

Vertical farming allows natural resources to

return to their natural habitat.

Historically, mankind has used land and water abundantly, with little regard for conservation. However, in recent times we have witnessed the harmful effects of meddling with nature.

Vertical farming allows farms to use a small space to build the farm. The plants are grown in a skyscraper built on that piece of land. This allows the Government to save thousands of acres of land. If this land is left as is, it will return to its natural state, allowing wildlife and wild plants to grow freely.

The demilitarized zone between North and South Korea is a good example. Before the Korean war, this area had a thriving farming community. However, ever since the area became a No Man's Land, it has been left to itself. In all this time, wildlife has rebounded, and today the area has a large population of Asiatic black bear, musk deer, and redwood crane.

- **WATER CONSERVATION**

Similarly, vertical farming allows conservation of water easy. Both hydroponics and aquaponics recycle used water. While aeroponics requires only a small amount of water to be sprayed on the roots.

All three of these methods require far less

water than conventional farming.

- **SCARCE RESOURCE**

Resources, like fertilizers, are abundantly used in traditional farming methods. However, the raw material for these commodities is limited and might run out sooner or later.

Vertical farming uses a closed-cycle design which limits the depletion of nutrient solution.

- **MASS EXTINCTION**

The current wave of mass extinction of animals is a direct result of large scale agricultural activities.

In many third world countries, when large pieces of forests were cleared, the animals living within those forests were also killed. Vertical farming prevents mass extinction by returning land and the animals on it to return to their natural habitat.

- **HEALTH**

Farmers who work in traditional farming fields are usually exposed to various infections.

These infections can range from malaria to exposure to pesticides, or even deadly attacks from venomous snakes. Farmers might also get hurt while using heavy farming equipment. Vertical farming allows farmers to grow food in a safe place.

It also allows food to be grown and supplied cheaply.

- **URBAN GROWTH**

As most people migrate to cities, Governments around the world are left to deal with a possible food shortage. Large cities have always imported food from other parts of the world. However, as more and more people migrate from rural areas to urban areas, the agriculture industry faces a shortage of human labor. Hence, for cities facing this problem, vertical farming is a blessing. It allows cities to become self-sufficient in food production.

- **RENEWABLE ENERGY**

It is known that vertical farms, or any farm for that matter, produce greenhouse gases. These are a mixture of gases produced from organic farming waste. Greenhouse gases are water vapor, carbon dioxide, nitrous oxide, and methane. Among these, methane is commonly used as a source for renewable energy. Farm owners can have methane digesters built on site. Methane digesters will generate electricity for the farm allowing it to save the money it had to pay the electricity company.

- **EMPLOYMENT OPPORTUNITIES**

Conventional farming is usually conducted by semi-educated farmers.

There is little professional education these farmers need to operate a farm. However, vertical farms are run entirely by educated people, who consider agriculture a scientific field.

Since a vertical farm is built in a city, it is a good chance to give jobs to these educated young people. Apparently, this will help decrease the unemployment rate.

- **WASTE MANAGEMENT**

Today, the vast majority of people use advanced technologies which the previous generation could not even think of. The human nature of solving problems has led us to a point where we can perform tasks without even getting up from our chair. However, this has also created various problems that were meaningless a few years ago.

Waste management has become a serious issue today.

Both, the developed and developing nations have devised different ways to solve this problem. Though, despite these methods, the problem still exists. Up till now, most countries have treated waste by having it compacted, and left in landfills to rot. Some countries produce burn waste to produce electricity. Liquid waste, on the other hand, is treated with chlorine and dumped in the river, or the sea.

All of these methods solve the problem of discarding waste. Though, they do not give a permanent solution. The waste continues to exist, even if it is removed from one place. Experts claim that vertical farming has the potential to solve the waste problem.

Apparently, several forms of waste can act as a form of nutrient for the plants.

Take organic waste, it mainly comes from the restaurant industry. Organic waste is a good source of methane. It has been discussed in the previous section that methane gas could be used by vertical farms to produce electricity. In fact, methane gas can provide such a significant amount of electricity that vertical farm owners won't have to buy electricity from the grid, and at the same time, they would be helping the restaurant industry get rid of infested garbage for free.

Another type of waste is polluted water. Currently, developed countries treat polluted water by mixing it with chlorine and then dumping into any body of water. This treatment partly reduces the harmful chemicals, present in the water. While developing countries don't even bother to do this. They simply dump the dirty water as is, into the sea.

Apparently, both of these methods cannot be considered a permanent solution. Both the solutions leave harmful agents in the water. Vertical farming, on the other hand, provides a far better option. Agricultural scientists believe that polluted water, from sewage, serves as a nutrient for plants.

Hence, vertical farms can extract methane from the water, and use the water to serve as a nutrient solution. The methane will be used to generate electricity for the farm, while the water will be passed through the hydroponics system.

If the farm grows certain plants like; Water Lilies, Cattails, and Moss, they can absorb the harmful components from the contaminated water, cleaning it in the process.

• NO POLLINATION WITHOUT INSECTS

Both professional farmers and agricultural scientists know that pollination is vital for a plant to grow. Like all living creatures, flowering plants too, have different organs; or parts, that are vital for reproduction.

The reproduction process of flowering plants is called pollination.

Flowers have both male and female parts. The male parts, called stamen, produce a sticky powder which is known as pollen. The female part is known as the pistil. At the top of the pistil, there is a sticky portion, called stigma. In order for pollination to take place, the pollen must be transferred from a stamen to the stigma. The interesting thing is that pollination only takes place if both the plants are of the same type. If the plants are of a different type, then it won't work.

It is understood that pollination helps flowers grow into fruits. Without it, we won't have any plants, or in broader terms, any food. However, the question is how does pollination take place?

Or in other words, how does pollen from one part of the plant move to another? Well, there are several different ways pollination can take place.

Humans can transfer pollen from stamen to stigma.

Though, nature does not require humans to interfere with its work.

The natural process involves insects, who act as pollinators. These pollinators can be honey bees, bumblebees, moths, or even a common fly. The interesting thing is that while pollination is a natural process, the insect itself thinks that the whole process is an accident. These insects have no intention of pollinating the plant.

Instead, they are there to get food for themselves. Yes, the sweet nectar is in fact a form of food for the insects. And while these insects are feeding on the nectar, they unknowingly rub their bodies against the stamen and get the pollen stuck all over them.

When this insect moves to another plant, the

pollen grains are transferred to that plant's stigma.

There are certain plants that have long stamens and pistils. These are pollinated by wind. When the wind blows, it picks up pollen from one plant and transfers it to another. Such plants don't need to attract animal pollinators, which is why these plants normally look dull, with small petals and no scent.

Plants in a vertical farm system are inside a building. This means that it is impossible for any insect; like the honey bee, to be there unless the farm management has a built-in beehive.

Additionally, there is absolutely no possibility of wind blowing inside the vertical farm. Hence, the only other available option left is to pollinate the flowers manually.

However, manual pollination is costly and time-consuming.

Vertical farms have plants stacked in layers, in a tall building, so the employees would have to spend a considerable time climbing up and down to pollinate each and every plant. This seems to be a useless expense.

- **ECONOMIC FEASIBILITY**

The main idea behind vertical farming was to reduce food miles i.e., the distance that is covered from food to end-user.

This idea was accepted by critics because

they believe that instead of buying food from a third world country; which forces poor farmers to work for low wages, it would be better to buy locally grown food.

Secondly, most critics claim that conventional farmers use chemicals to improve soil fertility and extend the plant's life. This harms the food and in turn causes harm to the consumer. However, since vertical farms grow food in a soil-less environment, inside a building, there is little chance that pests will get there.

Considering the facts mentioned above, most people ask whether vertical farms are capable of earning a profit.

All the environmental benefits of vertical farming can be considered useless if the business isn't generating any money. However, a recent report suggests that transportation accounts for only a minor portion of the costs of food distribution to cities. This means that vertical farms use distribution prices as a marketing fad, and they would have to sell the produce at a higher rate to earn a profit.

Vertical farms are normally built in a skyscraper.

Skyscrapers are not cheap. It takes a lot of money to build one skyscraper. On average, for a 60 hectare farm, the building cost can go higher than $100 million. And while building a vertical farm within the city might be a plus point, but it's not. Real estate prices within the city are too high for startups.

- **ENERGY USE**

For humans, sunlight produces Vitamin D, which helps human body grow.

In the same way, sunlight triggers a process called photosynthesis, which helps plants in their growth. Unlike conventional farming, where plants receive sunlight for free, vertical farms must provide artificial light to the plants. And for this, they must burn a lot of energy. According to George Monbiot; environmental activist, the cost of providing light, to grow one single loaf, would be $15.

Vertical farms are highly dependent on energy. From electricity to climate control, everything consumes a lot of electricity. We discussed previously that vertical farms can build methane digesters that will power the entire operation. But then again, it is an additional cost which can have a significant effect on the operational cost.

- **POLITICAL IMPACT**

Vertical farming is a good concept for most countries. However, if cities become heavily reliant on it, they would no longer have to buy food from third world countries. Hence, farmers in these countries would be left unemployed.

Agriculture is the most important industry in most third world countries. The Governments have made significant investment in agriculture and cannot afford to lose sales.

The main drawback of third world countries is that most people who live there are either uneducated, or semi educated. And since occupations like farming do not require a degree, the people don't bother to educate themselves. Apparently, if their livelihood is affected, then they won't have any other alternative.

The mass unemployment in third world countries, caused by vertical farming will have far reaching effects. It won't only affect economically; it will have political impact also. The unemployment might kick start a chain reaction, which might harm the bilateral relations of the two countries. Hence, it is quite apparent that while vertical farm might be a good idea for the developed world, it might not be for the global economy.

- **POLLUTION**

Vertical farms consume a lot of electricity to power the lighting system.

We have mentioned in the previous section that electricity is produced from the greenhouse gases.

However, what remains a major concern is

the pollution caused while producing electricity. This pollution can be attributed to the greenhouse gases that are produced at a vertical farm.

Greenhouse gases comprise of methane, water vapor and carbon dioxide, among others.

Carbon dioxide is vital for plants, as it increases the rate of photosynthesis by almost 50%. There are some farms which prefer burning fossil fuels; because fossil fuels help produce a more refined version of carbon dioxide.

Carbon dioxide may not be the only pollutant. Another major pollutant is the nutrient solution, commonly used in hydroponic systems.

We mentioned in a previous section that farms use plants to clean the nutrient containing water. We also mentioned that this water is reused after its cleaned naturally. However, it should be noted that after being used more than once, the nutrient solution reaches the limit and has to be disposed off.

The problem is, that solution contains chemicals; such as fertilizers and pesticides, which can be harmful for the environment. Since vertical farms are based inside cities, it would be difficult to just throw the water out, unlike conventional farms.

- **AN EXPENSIVE HIGHLY TRAINED WORKFORCE**

Vertical farming, unlike conventional farming, requires educated and trained professionals. Professionals who consider this system a science, not mere farming. Unfortunately unlike conventional farmers, engineers, horticulturists, agriculturists and technicians; who live in cities, charge a lot of money. These professionals can claim that they have certificates, diplomas and degrees from universities; this proves that they are qualified for the work. Add to that the years of experience they might have.

Most vertical farms still make their employees conduct certain tasks manually.

For example, pollination is still done manually, because the farms do not have any insects to replace the manual labor. Automating the establishment is an option, but a very costly one. Considering the costs of hiring such a large, educated and experienced workforce it is quite apparent that costs will be in millions of dollars.

Most traditional farms use poverty to their advantage. Poor and illiterate farmers are made to work for minimum wages. These farms are often based in a third world country where labor laws are not enforced as such. Hence, a harvest would normally cost a farming establishment lesser than it should. Consider this; Starbucks buys its coffee beans for $1.25, in order to serve it for $4. So for $1.25, those coffee beans are transferred from an African coffee plantation all the way to Starbucks. On the other hand, if a vertical farm grows coffee it will definitely cost more than that.

- **ONLY A LIMITED TYPE OF CROPS CAN BE GROWN**

Since vertical farming is an expensive business, farmers have to grow crops that can help the establishment rake more profits. Hence, for the time being, vertical farms can only produce crops that have a high value and can yield rapidly. Crops that can be harvested many times in a single year are ideal for vertical farms to cover their expenses.

Plants, commonly used in salad, and those that have medicinal values are ideal for vertical farms. These plants can grow rapidly and over the year, they can be harvested many times. Besides, they also have small footprints and won't cover much space. Slow-growing plants are not profitable as they take much more space and time to ripen. They might not make any significant profit for commercial farms.

GREEN BUSINESS

Vertical farming was once a revolutionary idea that mocked the traditional farming methods. Most traditional farmers thought that vertical farming was merely science fiction. Apparently, NASA's fascination with aeroponics system did in fact make it look like science fiction. However, fast forward to the 21st century, millennial entrepreneurs are willing to launch agricultural business which focus on growing food round the year.

Today, there are many people who no longer consider vertical farming a fantasy. There are many advocates who believe modern urban centers need vertical farms are a necessity.

The question that needs to be asked is; what made the educated class change their opinions on vertical farming? Apparently, when vertical farming was first introduced, people did not face the problems they face today. Back then, most of the people had more than enough to eat.

Nobody ever thought that with so many

countries contributing to the world's agriculture sector; we will run out of food. However, in recent times Earth's population has increased manifold.

We now have more than 7 billion people living on this planet.

The fast growing population and limited fertile land is a good reason to invest in vertical farming.

Investors now consider vertical farms a profitable investment opportunity. However, they investors didn't always feel this way.

Previously, most critics felt that vertical farming was too expensive to deliver any profits. But in recent times, cheaper versions of the tech gadgets have been introduced, which makes the vertical farming business seem more enviable than before.

The latest tech gadgets designed for vertical farming are perception technology; which has infrared cameras. These infrared cameras can zero on plants that are affected with diseases.

Then we have artificial intelligence; which is a good tool to process the information received from the perception tech gadgets. And finally we have automated robotics and drones. The drones can be used to extricate the specific plant which is infected, or spray it with fertilizer. The robots, on the other hand can be used to pick the produce when it is ready, and move it to packaging area.

There is a big difference between vertical farming and greenhouse farming.

Most critics claim that since both vertical farms and greenhouse, grow food indoors, they are both the same. However, that is the only similarity in the two systems. Research articles claim that greenhouses are dependent on sunlight, while vertical farms are run on artificial lights.

Secondly, greenhouses produce far less than vertical farms; which grow tons of plants stacked over one another.

Since more production means more profits, vertical farms can generate more profits for the owners.

As far as costs are concerned, greenhouses are more dependent on human labor than vertical farms. Vertical farms rely on cutting edge technology to limit the role of human labor and increase efficiency.

Despite the technological advancement, there are two factors which do not work in vertical farm's favor.

The first factor is; accessing land. It is a known fact that land is much cheaper in rural areas than it is in urban areas. In cities, open spaces are bundled in small plots that are sold for commercial or residential purposes. Whereas vertical farms require a large space in acres, to grow a significant amount of crops.

Another factor which applies is the potential to scale up operations.

Conventional farms have the ability to grow tons of food.

Most smaller farms fulfill the market demand by growing different types of crops on a rotation basis. Vertical farms, on the other hand, have a limited space. They cannot grow as much food. Firstly, there are very few types of crops which can be grown at a vertical farm. Cash crops like Wheat, rice and potatoes cannot be grown, as they are slow growing crops and might not earn a profit for the farm.

Since planting slow growing crops will be costly, vertical farms grow fast growing salad greens; like lettuce, to make a profit. However, if too many farms concentrate on lettuce only then it is obvious that the price of lettuce will fall.

It is quite obvious that our planet cannot be fed on lettuce alone. We need to grow a variety of plants at the vertical farms. Various reports claim that besides lettuce, eggplants, tomatoes, herbs and microgreens can also be grown. It has been reported that a vertical farm in France grows berries, as the French are fond of them.

The recently passed United States Farm Bill lays importance of indoor farming and emphasizes on researching new crops that can be grown.

Somehow it is apparent how the profitability problem will be solved.

We know that if we grow other crops; like spices, fruits, or medical marijuana, we might be able to solve the problem. Hence, it is up to the entrepreneur to decide whether he should stick to growing salad greens or do a bit of research and find out the perfect crop to make him rich.

CONCLUSION

If the United States Government has passed a bill, which mentions vertical farming, it is apparent that this is not a small project.

Vertical farming is a fast growing industry, with a potential to become a billion dollar industry in the next 5 years.

Many big corporations; like Google, IKEA, etc. have already invested millions of dollars in this venture.

The only problem investors are facing so far, that they are unable to find a cheap way to grow tons of crops.

The technology used is extremely expensive. Then again early stage technology is normally very expensive. Perhaps if we outsource the manufacture of agri tech gadgets to the third world, we might be able to solve the problem. At the same time places where sunlight is abundant; like California, can take advantage of it. They can use the sunlight to power their LED lights.

We all know that there are a variety of crops that can be grown in a vertical farm. The only reason why farmers grow lettuce is because it has a shorter cycle. In order to make money, we need to branch out and grow other crops. Otherwise, we might only end up flooding the market with cheap lettuce.

The idea behind vertical farming was to produce plants that are fresh, free of fertilizers, help reduce the carbon footprint and profitable for the business.

These factors have already been achieved by most of the farms. Though they are still worried about profitability. It is a widely known fact that early stage technology is expensive.

Although, more research can help businesses come up with a way to produce more within the budget.

If you look at the bigger picture, you will realize that vertical farming is the future.

In the future only slow growing crops; like wheat and rice, will be imported from the third world. It is possible that big restaurant chains might have their own vertical farms one day. That way they will become self-reliant.

5 years from today, the people who are investing in vertical farms now, will be considered pioneers. And it is highly likely that

our research will be taught at universities. It is up to you to decide, whether you want to be called a pioneer, or a late-starter.